" 16 God made the two great lights: the greater light to rule the day, and the lesser light to rule the night."
(World Messianic Bible, Genesis)

When we look up at the sky, there is the sun shining brilliantly. It dawns with the light of the sun to create a day in the night. Bathed in the sunshine, living things in nature start throbbing pulse on the earth. Green plants grow and flower, wild animals run around on the ground, and birds fly in the sky. Clouds grow up in an air current ascending from the sea, and cumulonimbus causes rains and storms. The sunshine also creates a hurricane at the summer ocean, which travels fast from the sea to the land.

The sun shines on the moon to create waxing and waning, which brought the solar and lunar calendars during a long history of human beings. The sun has long been an object for natural worship as the Sun God "Ra" in ancient Egypt, as Sun Goddess "Amaterasu" in ancient Japan, and also as Sun God "Dainichi" in Buddhism. The earth, on which we live, travels around the sun together with other seven planets to form a system called the solar system.

The sun is one of billions of stars in the Galaxy. It is important for us to know the nature and history of the sun along with the evolution of the solar system in order to prepare for the coming space period. Especially, the sun is a typical star in the Galaxy and thus should give us a clue to know the nature of stars twinkling so high in the sky.

In this book, I would like to consider the nature of the sun and the history of the solar system from a standpoint of natural philosophy.

At home in Kasukabe

Jun 8h, 2015

Hiroyuki Aizawa

Contents

Chapter One
Nature of the Sun

The sun rises in the east and sets in the west traveling in the blue sky, creating day and night. In a solar calendar, the cycle of rotation of the earth around the sun and on its own determines a year and a day, respectively. Our modern solar calendar also takes a cycle of revolution of the moon around the earth as a month. The solar calendar includes almost all the essences of nature of the solar system, which has been revealed by classical physicists including Galileo, Descartes, and Newton. Among them, the Copernican theory is one of the most important secrets of modern astronomy. After the discovery of the theory, astronomers and religionists took a thorny path to establish the solar calendar with a long history of wars, into which a lot of politicians, soldiers, and people were dragged. We could say that the common knowledge of the solar system is a fruit of an effort in the long history of human beings.

The sun is the largest and heaviest spherical body in the solar system, and the sun locates at the center of the system. The diameter, volume, and mass of the sun are estimated to be a hundred times longer, 1.3 million times bigger, and three hundred thousands times heavier than those of the earth, respectively. The sun is a huge body, which occupies more than 99% of the total mass of the solar system. And the sun is the only star, which radiates light, in the solar system. Its surface temperature appears six thousands centigrade degree, and could reach up to twenty million centigrade degree at the flare of a corona. The total radiation power of the sun is estimated to be 400 Yotta-Watt, which corresponds to four thousand billions of one Giga-Watt nuclear power plants.

Sometimes we find some sunspots on the surface of the sun. The sunspots show various sizes and shapes one by one, and an average of life span of sunspots is around two weeks. Astronomers studied on rotation of the sun by precise observation of sunspots. As a result of the observation, it has been revealed that the cycle of rotation of the sun is 27 days. The axis and direction of rotation of the sun are almost the same as those of revolution of planets around the sun, respectively.

The sun is a star in the solar system, and is also one of a lot of stars in the Milky Galaxy. The Galaxy is speculated to be a typical spiral galaxy, which contains several spiral arms extending from the center. Each arm consists of billions of stars, and the sun belongs to one of the spiral arms. The Milky Galaxy consists of a central bar-shaped bulge and a peripheral flat rotating disc just like an egg fried sunny-side-up in its shape. The sun locates at the middle of one of the spiral arms in the flat rotating disc of the Galaxy.

How was the sun born during an evolution of the Milky Galaxy?　Evolution of the Galaxy is an awesome and magnificent story, but it goes far beyond the scope of this book.　Here in this book, I would talk about the history of the sun from its birth to the present along with other planets in the solar system.

Chapter Two
Nature of Planets

We have eight planets revolving around the sun. Among them, we could see Mercury, Venus, Mars, Jupiter, and Saturn by naked eyes, and ancient astronomers knew well about those five planets. In ancient days, it was believed that the universe is made from five elements, that is, water, metal, fire, wood, and soil, each of which were derived from the five planets, Mercury, Venus, Mars, Jupiter, and Saturn, respectively. This is The Theory of Five Elements in the ancient china. To tell the truth, the relationship between the solar planets and the ultimate elements is quite important to understand the history and whether of the solar system as described in detail at the chapter seven. Seven days of the week consist of the five planets and the sun and the moon basically from the theory of Yin-Yang and the five elements, where Yin and Yang correspond to the moon and the sun, respectively. In this calendar, four weeks or 28 days correspond to the cycle of waxing and waning of the moon. Accordingly, the four weeks became a unit of a month as a basic structure of the lunar calendar. The motion of planets in the solar system has always been playing a central role in setting a calendar as a natural clock for days, weeks, months, and years.

On the other hand astronomers found two more planets, Uranus and Neptune, between late eighteenth century and early nineteenth century after the development of high-resolution telescopes. Each of the eight planets revolves around the sun along its own circle orbital. The closest planet to the sun is Mercury. The average density of Mercury is as high as 5 g/ml while its total mass is as small as 6 % of that of the earth. The diameter of an orbit circle of Mercury is less than 40% of that of the earth, and it takes 3 months for Mercury to revolve around the sun. The next planet close to the sun is Venus. The average density of this planet is also as high as 5 g/ml, and its total mass is around 80% of that of the earth. The diameter of an orbit circle of Venus is around 70% of that of the earth, and it takes 7 months for Venus to revolve around the sun.

The next planet close to the sun is the earth. The average density of the earth is as high as 5 g/ml, and the earth is mainly made of iron. The earth revolves around the sun once a year. Since the earth is a huge magnet, it is speculated that heated iron solution was cooled down to become solid under a magnetic field during the creation of earth in the history of solar system. The earth has a single satellite or the moon, which rotates around the earth and itself exactly at the same direction and periodicity with a 28-day cycle. Composition of stones of the moon is almost identical to that on the earth, while there is little atmosphere and water on the surface of the moon. Diameter of the moon is one forth of that of the earth, and the total mass of the moon is only 2% of that of the earth. Consequently, the gravity on the moon is one sixth of that on the earth, which may cause the loss of atmosphere and water from the surface of the moon. Since the surface of the moon is naked of water and atmosphere, almost all the meteorites attack directly on the surface resulting in lots of huge craters. We see the craters, which appear as rainfall patterns on the sandy ground, from the earth using a commercial telescope.

The next planet close to the sun is Mars. Nowadays, an astronomic search for the Mars became a hot topic. Especially, the existence of living things on the Mars now and then is one of the greatest concerns in the modern science, because the Mars appears quite similar to the earth. It is expected for astronomers to discover any evidences of life on planets other than the earth. Mars appears a little bit smaller planet than the earth, and the diameter, volume, and mass of the Mars are around a half, one seventh, and one tenth of those of the earth, respectively. The diameter of revolution of Mars around the sun is one and a half of that of the earth, and the cycle of the revolution of Mars is about two years. The Mars revolves along the most outer orbit among the four terrestrial planets of the solar system, and billions of tiny small planets called an asteroid belt revolve just outside of the orbit of Mars. Each asteroid such as Itokawa is not a spherical body and takes a unique irregular shape. These steroids might be underway to assemble with each other by universal gravitation to create a new planet between Mars and Jupiter. It is also speculated that the steroids are the wreckage of a planet that was destroyed long time ago. Those hypotheses should be verified by a precise observation of the behavior of asteroids in the belt. We will discuss the creation and history of the sun and planets in detail in the latter half of this book.

Planets revolving outside of Mars consist of relatively light materials such as water and low molecular mass organic compounds. Jupiter, the most inner planet among them, has an average density of 1 g/ml, while its diameter, volume, and total mass are eleven, one thousand, and three hundred-times bigger than those of the earth, respectively. The diameter of the revolution of Jupiter around the sun is five times longer than that of the earth, and the cycle of the revolution is about twelve years. Saturn is a huge planet with a ring-shaped belt of icy satellites around it. Its diameter is 10% smaller than that of Jupiter, and its volume and mass is eight hundred and one hundred times bigger than those of the earth, respectively. On the other hand, its density is the smallest among those of solar planets and less than 1 g/ml. The diameter of revolution of Saturn around the sun is ten times longer than that of the earth, and the cycle for revolution is thirty years.

Diameters, volumes and masses of ice giants, Uranus and Neptune, are four, sixty, and fifteen times bigger than those of the earth, respectively, and the two planets resemble to each other in its shape, mass, and composition. The diameters of revolution of Uranus and Neptune around the sun are twenty and thirty times larger than that of the earth, respectively. The revolution cycles of Uranus and Neptune are eighty years and hundred and sixty years, respectively. It is also known that dwarf planets such as Pluto revolve around the sun outside of the Neptune. Outside of the dwarf planets, comet's hometown or the Oort cloud is speculated to exist as the outer limit of the solar system. The diameter of the revolution of the Oort cloud is estimated to be more than thousand times larger than that of the earth.

As we discussed above, the solar system contains eight planets, millions of asteroids between the earth and Mars, dwarf planets outside of the Neptune, and comets in the Oort cloud. How were those bodies created in the history of the solar system? Were they created altogether as sons of the sun? Or were they born and grown independently at different places in different periods and then joined to the solar system one by one? Do they revolve around the sun in an independent manner to each other? Or do they revolve around the sun cooperatively by some power or connection with each other? Let us start to discuss the genesis and history of the solar system in detail from now on.

Chapter Three
Motion of the Sun and Planets

The solar system consists of the sun, eight planets, many satellites, an asteroid belt, dwarf planets, comets in the Oort cloud, and gas among the bodies in space. All of them are made from elements, showing various characters according to the species and combination pattern of atoms. They also perform dynamic motion according to the kinetics. Here, let us take notice on the motion of spherical bodies in the solar system.

There are two fundamental motions of spherical bodies in space, that is, a linear motion and an elliptical motion. In the solar system, all the spherical bodies take an almost circular elliptical motion for revolution around the sun. This suggests that the solar system is maintained by a centripetal force derived from universal gravity of the sun as suggested by Sir Isaac Newton in the eighteenth century for the first time.

There are two types of revolutionary movement for a spherical body. One is rotation and the other is revolution. The former is an internal motion of the spherical body, whose center of gravity does not move during the motion. The spherical body rotates around the axis of rotation, which passes the center of gravity of the body. We could set a rotation vector on the rotation axis according to direction and angular velocity of the rotation. When we put right hand fingers along the movement of the rotation, the thumb indicates the direction of the rotation vector. The length of the vector corresponds to the double of the angular velocity of the rotation. For example, the rotation vector of the earth directs from the South Pole to the North Pole, and its length is one sixth of π radian per hour.

On the other hand, revolution is a circular movement of a body around a certain point in space. In the solar system, planets revolve around the sun. We can set a rotation axis and a rotation vector of revolution as in a case of the rotation. The revolution of each planet is performed under the law of classical physics by the balance of centrifugal force and universal gravitation. Mechanics speculates a precise orbital path of each planet by computational calculations. We will discuss the method for orbital calculation in detail in chapter five.

The rotation and revolution of the sun and its eight planets clearly indicate an interesting tendency at a glance. Each planet revolves around the sun on a concentric circle with different diameter. Amazingly, all the circles of the planets locate almost in a single plane. Furthermore, all the planets revolve in the same direction. In other words, all the rotation vectors of eight solar planets point to the same direction with different size.

The rotation vector of the sun also points to the same direction as those of eight planets. Since mass of the sun corresponds to more than 99% of total mass of the solar system, the coincidence of the direction of rotation vectors of the sun and revolution vectors of eight planets suggests that the solar system itself rotates coordinately around its center of mass as a whole. While the sun and planets are independent spheres, they rotate as if they are parts of a whole body. This result indicates that the solar system behaved as a single matter just upon the birth. We will discuss this issue in detail in the next chapter.

Let us see the rotation vector of planets around itself. The direction of rotation vector of the sun is almost the same as that of revolution vector of planets. However, Venus rotates to an opposite direction and rotation vectors of the earth and Mars incline at an angle of 25 degrees from their revolution vectors respectively. Jupiter rotates around an axis of revolution in the same direction, while Saturn, Uranus, and Neptune rotate around their axes, which direct almost randomly. Taken altogether, we could not find any rule in the direction of rotation vectors of planets. We could not find any relationship between rotation vector and rotation radius of planets, neither. On the other hand, we find an obvious relationship between revolution vector and revolution radius of planets, from which Sir Isaac Newton discovered the law of universal gravity.

In this chapter, we have investigated the motion of spherical bodies in the solar system in detail. We recognized a rule for the plane and direction of revolution of the planets without exceptions, suggesting secrets of the birth of solar system. When compared to the perfect coincident of the revolution plane of planets, random direction of rotation vectors of planets even appears funny. We could tell two possibilities to explain the random rotation of planets.

One possibility is that some magnetic force perturbed the rotation of planets. When a droplet of iron-rich metal liquid is slowly cooled down to form a solid planet in space under a magnetic field, the planet becomes a permanent magnet. If the magnetic field altered after creation of the planet, the planet should shift the direction of rotation by the torque of the magnetic force. This hypothesis might be applicable to planets rich in iron metal such as the earth.

Another possibility is the perturbation of rotation by a direct attack by comets or meteorites. It is likely that all the rotation vectors of eight planets point toward the same direction as those of revolution vectors at the genesis of the solar system. After the establishment of the orbital structure in the solar system, large comets and meteorites came down from outer orbits to the planets, and sometimes they attacked directly the planets to tilt the axis of rotation in a random manner. In this hypothesis, the attack affects not only the angle of rotation axis but also the angular momentum. It is no doubt that a lot of meteorites collided with the moon to make craters on the surface of the moon. The collision with comets or meteorites affects revolution as well as rotation of the planet, while the effect on the revolution is much smaller than that on the rotation because of much larger angular momentum of revolution than that of rotation.

Those two hypotheses may explain general aspects of behavior of the solar system quite well although we do not yet have precise concrete evidences, which tilted the angular of rotation axis of each planet along its history. For example, if the rotation axis of the earth tilts from that of revolution by magnetic force of the Galaxy, all the other magnetized bodies in the solar system should have the same tendency as the earth. It might be difficult to get evidence of the magnetic effects on the axis of planet rotation in the past. It is necessary to develop the solar science to evaluate the effects of size, mass, speed, and angle of meteorites on the linear movement, rotation, revolution of the planets.   We also need to confirm the effect of collisions by concrete evidences obtained from geological investigation. Hopefully, we will success to speculate and measure the tilt of rotation and revolution of planets by gyro effects of collision with meteorites in the future.

Chapter Four
Vortex Aspect of Planetary Orbits

As we discussed in the chapter three, all the revolution vectors of eight planets in the solar system point to almost the same direction as the rotation vector of the sun, whose mass is 99.9% of that of the solar system, indicating that the solar system appears to rotate around itself like a flying saucer. However, we do not detect any material structure connecting the sun and planets with one another. On the contrary, the outer planet revolves with longer cycle, suggesting that each planet revolves independently under the rule of Newtonian mechanics. I guess that those independent planets and the sun had previously been assembled into a single body as a rotating flying disk.

We can find some example for rotating disk-like body in nature. A flying disk is a familiar example for rotating saucer to the owner of a dog. For another example, let us think about how to make a round pizza from viscous wheat dough. A pizza chef stretches the dough by centrifugal force into a perfect disk by whirling it in the air. Just like the pie dough, a drop of liquid water may become a disk-like shape when it whirls in the sky. Another example is a hurricane in the Atlantic or a typhoon in the Pacific, which develops a large disk-like maelstrom of clouds, which is made from a vortex steam in an ascending current from surface of the warm sea.

Although eight planets revolved around the sun at present, it is quite likely that the solar system was born as a body of vortex vapor consisting of all elements in chaos. Probably, the solar system was created as a body of high temperature, high pressure, and high-density gas, which was sent out in space at birth. It is speculated that the gas might expand quickly into space with decreasing its temperature, pressure, and density after the creation. The gaseous body emitted heat into space, and consequently the gas started shrinking and vortex rotation. As it shrank more, it rotated more quickly. Since gas changes its shape more easily by centrifugal force than liquid, the gas changed its shape from a sphere to a disk during its condensation. During the shrinkage, aliquots of gas occasionally condensed into tiny liquid droplets that stay on the disk region even after almost all the residual gas condensed into a single spherical liquid body at the center of the solar system, that is, the sun.

All the liquid droplets emitted heats gradually, and consequently solidified to form a spherical body or a planet, which revolves around the sun within a plane of the vortex disk. Even after the separation, all the spherical bodies in the solar system keep the angular momentum of the original vortex vapor. This is exactly the Cartesian vortex theory proposed in the seventeenth century for the first time. The theory does not explain the relationship between cycle and diameter of the revolution of planets at present. On the other hand, Cartesian vortex theory explains how to make the solar system, and the theory straightforwardly suggests the causal relationship between the vortex movement of an ancient chaotic gas body and the present axes and directions of rotation movements of all the eight planets and the sun.

Chapter Five
Universal Gravity and Planetary Orbits

Even though the genesis of the solar system started from chaotic vortex vapor, the original vortex motion does not drive the present planetary motion any more because almost all the materials in chaos has been condensed and solidified in spherical bodies to make an ordered solar system. Also the vortex motion transferred to the automatic revolution and rotation of spherical bodies, which still preserves the primary angular momentum and kinetic energy in the system. In the eighteenth century, Sir Newton clearly discovered the universal gravity, which attracts planets to the sun by power inversely proportional to square of its orbital radius, from the well-known rule that a square of planetary angular velocity is in inverse proportion to a cubic of orbital radius of a planet. The universal gravity is the secret of planetary revolution and causes everything to fall on the ground.

As we learnt in classical physics class at high school, planets revolve around the sun on elliptic orbits accelerated by universal gravity. Planetary revolution is almost completely speculated by celestial mechanics while N-body problem taking account of interplanetary gravitation could not be solved strictly by mathematic calculation. We usually estimate a planetary revolution taking only the sun and a certain planet in account.

On the other hand, the universal gravity does not explain how all the planets revolve in a single plane to the same direction. If each planet joined the solar system accidentally one by one to get into an orbital path around the sun, it is almost impossible that all the planetary orbits set in a single plane. The solar system was created as chaos of vortex vapor of elements, and the vapor shrank and rotated to form a disk-shaped gas body by radiation cooling, and the disk of gas was further cooled to condense and solidify to form spherical bodies in a rotating disk of the gas. As a result, planets revolve around the sun in elliptic orbits by the universal gravity. The solar system was created by the harmony of vortex, condensation, and gravity.

Chapter Six
Mass Distribution of the Solar System

According to the evidences discussed above, the solar system has apparently behaved as a single substance from the genesis to the present period, while the rule governing the system altered occasionally during the course of development. The solar system was created as a vortex vapor, which condensed afterwards into liquid drops with radiation cooling. After the solidification of the droplets, the solar system is now organized as a sphere under the universal gravity of the sun. The solar system has developed continuously as a single object in space all through its history. The motion of the substance is represented as a mass distribution function of a mediator or time as a variable. The solar system is not the exception but having its own function of mass distribution although some materials might get into and leave away from the system during its evolution. While we need four dimensions to express time-dependent distribution of mass, here we show the three dimensional function at a certain time fixed at one of the three typical stages of the evolution in a simple and easy way. In this chapter, we aim to investigate on the nature of the function.

The newborn solar system is speculated to have been a

vortex vapor at high temperature, under high pressure, and with high density. Consequently, its mass distribution function should obey the ideal gas law. Since the surrounding space is almost a vacuum, the blowing steam should expand adiabatically. The primary vapor is a mixed gas containing all the elements in chaos. Furthermore, the mixed gas is stirred because the vapor was whirling around in space. If the mixed gas stays still in space at zero gravity, high-density heavy metal gas accumulates at the center of the solar system as a core, and light metal gas may surround the core, and a low atomic mass element gas surrounds outside of the metal gas. All the elements should take their position from center to periphery of the gas body exactly in the order of the atomic mass. The element layers are governed by gravity within its sphere, and the gas could not retain any materials released from the gravity. However, the primary vapor is created under high pressure when compared to the outer space. Therefore, the gas is expected to expand quickly and radially after the birth. In this model, the primary temperature, pressure, and density of the vapor and the law of gravity determine the total mass and volume of the solar system. Moreover, the vapor was billowing here and there, and thus it is more complicated to calculate the total mass within a gravity sphere than that of a still gas. The billows of steam may also disturb the layer structure of the gas, to make calculation of gravity complex. Moreover, the composition of elements is expected to change dramatically after the birth of the solar system because of nuclear fission of radioactive heavy elements, which should be rich in the primary vapor. And, adiabatically expanding cooling of the vapor caused chemical reaction of the gas to affect the number of molecules, the compositions of the gas, and the density. Taken all together, it is almost impossible to describe and speculate in detail what is going on in the primary gas of the solar system. It is really chaos. However, there is little doubt that a primary vortex vapor adiabatically expanded into space with decreasing its density and temperature just after the birth of the solar system.

Next, let us think about the mass distribution of the solar system during the condensation. How is the shape of dewdrops condensed from the primary vapor in space by radiation cooling? Please imagine boiling water in a pod. When steam is blown out of the spout of teakettle, hundred degrees vapor condensed into droplets in air with its volume reducing to one per two thousands. On the assumption that temperature of the primary vapor is a hundred million kelvins, a volume of the vapor reduced to one per twenty thousands at five thousand kelvins, which is a condensation point of heavy metals. And it further reduced to one per a thousand after the condensation to become a drop of liquid metals. Since the angular momentum of the body is conserved during the condensation, the body rotated fast as it shrank, resulting in a radial centrifugal force in a rotation plane but not along the rotation axis. Consequently, the vortex vapor shrinks more quickly along the axis direction rather than in the rotation plane to form a disk-like structure. When cooled to the condensation point, each element in the vapor disk suddenly starts to condense into a liquid droplet at the center of the vortex. Almost all the mass of the element condensed at the center of the vortex to form a large liquid droplet as a seed of the sun while some residual vapors condensed here and there to form liquid droplets as planetary seeds in the disk region. Almost all the droplets with relatively slow velocity were attracted to the center by gravity and join the large droplet to form the sun. Only several droplets with exceptionally fast velocity could ride on orbits around the sun only when their centrifugal forces balanced with the gravitational ones. Even if the original vortex motion of the primary vapor was just a small tremor of the gas within the Oort clouds, the angular velocity of the rotation of the sun and revolution of planets should increase to some extent after the condensation because of the law of preservation of the angular momentum. In this way, 99.9% of mass finally condensed in the center as the sun while a lot of spherical bodies including eight planets revolved around it in the same angular direction. And so a prototype of the solar system formed.

How does a condensed liquid droplet look like? Is it spindle-shaped, spherical, or drop-shaped? In space, there is no imbalanced force like surface tension and gravity on the drop-shaped water dripping from a tap. Consequently, the condensed liquid drops of the solar system are expected to be almost spherical. The sun rotates 15 degrees per day around itself at present. Such a slow rotation may not affect the shape of the spherical body so much if the body is rigid solid, elastic solid, or highly viscose liquid.

Lastly, let us think about the mass distribution function of the present solar system. When compared to the drastic change of the shape, states, and density on the condensation, the present solar system appears stable to some extent. Liquid droplets of the sun and planets solidified at freezing points of each element by radiation cooling. The perfect spherical shape of the sun and planets at present should be the consequence of the solidification of them in space. Assembly of meteorites by collision does not result in a spherical shape of the sun and planets, and also does not result in the revolution of eight planets around the sun in the same angular direction. In the sun and planets, heavy and light elements locate inside and outside of the spherical body, respectively, and heavy and light gas surround the spherical body near and far from its surface, and the lightest gas such as hydrogen and helium fill the interplanetary space in the solar system. This is the mass distribution function of the solar system at the present.

Chapter Seven
Ultimate Analysis of the Solar System

Modern western science revealed that the world consists of ninety-two elements. Almost all the elements around us on the earth are stable isotopes with little radiation having very long half-life compared to the age of the earth. On the other hand, there are various radioactive isotopes such as uran-235 with a half-life of a billion years, which is comparable to the age of the earth. We already discussed the solar system from the birth to the present standing on a view of physical nature such as mass, volume, and temperature. In this chapter, let us stand on another point to see the history of the solar system from a view of distribution of elements, which determine chemical nature of a substance.

What kind of elements does the solar system had upon the birth as a vapor? Although we are not sure about history of the world far before the birth of the earth, we can assume that a content of high atomic mass elements was higher in the primary vapor than that in the present solar system. Radioactive isotopes disintegrate into small elements by nuclear fission at their specific half-life and the radioactive ray also induces the collapse of atoms into hydrogen. Therefore, we speculate that the primary vapor contained higher content of heavy elements than the present solar system while I could not tell the exact distribution and content of each element of the vapor in chaos. After the condensation of the vapor into liquid droplets, elements performed a characteristic distribution pattern in the solar system. Upon a condensation of a vapor of various elements in chaos, each element or molecule is sequentially distilled and purified in order as distillation of the whiskey. When a gas of mixture is cooled down gradually, materials liquefies in order of the condensation point from high to low. If a vapor mixed with water and alcohol is gradually cooled down at 1atm, water condenses at 100 degrees and then alcohol condenses at 80 degrees. In case of the primary vapor of the solar system, heavy metals such as tungsten is estimated to condense at around 5000 degrees, and then iron at 3000 degrees, cupper at 2500 degrees, silver at 2000 degrees, and mercury at 360 degrees at 1atm. Probably, elements with a high condensation point forms nuclei of the condensation to form the sun and planets, and the nuclei grew incorporating elements with lower condensation points in order. As the sphere grows the pressure in the body getting higher by gravity of its own, resulting in higher condensation point. The surface temperature of the present sun is 6000 degrees, at which almost all the metals vaporize at 1atm. However, the pressure is so high by gravity that the metals are expected to liquefy or solidify inside the sun, which is surrounded by gas of light elements such as helium under high pressure.

Each planet is made from 0.01% of total materials of the solar system by condensation of the primary gaseous chaos of elements independently. Planets are much smaller than the sun, probably because partial and quick cooling of the gas resulted in a tiny drop of liquids or because materials with lower condensation point condensed slower and later during the cooling down of the primary hot gas for the solar system genesis. Ultimate analysis of each planet may tell us some aspects of the primary gas in the period and at the region, when and where the planet was created.

In the planetary science, relatively inner orbital planets, Mercury, Venus, Earth, and Mars, are called rocky planets, while outer planets, Mars and Saturn, are called Giant Gas, and the most outer planets, Uranus and Neptune, are called Ice Giant. These names of classification depend on the composition of elements in each planet. In other words, the composition differs from planet to planet according to the distance from the sun. The inner rocky planets are rich in metals and silicon dioxide. Outer planets or Gas Giants may contain some organic compounds such as amino acids and alkenes, although its ultimate analysis is still under investigation. The most distant planets from the sun contain a lot of water as ice together with low molecular mass chemicals such as methane and ammonium. The molecular mass of those light materials is around twenty.

Although we still have some limitations to access data of ultimate analysis of planets until now, it can be safely concluded that low molecular mass lighter gas materials stay outer region of the solar system while heavier gas materials locate inner region of the system. It may be caused by physical nature of the gas such as density and condensation point that low condensation point light materials tend to locate outer region in the solar system. How does such an orderly tendency of elementary composition happen in the solar system? It may be happened at birth of the solar system by ordered condensation or distillation of elements, and maintained by effects of the weather in the solar system as discussed in the chapter nine.

Chapter Eight
Genesis of the Sun

It was revealed that the sun was created as a star at the center of the solar system by condensation and congelation of the primary vapor with elements in chaos at high temperature under high pressure. When a volume of the vapor reduced dramatically upon the condensation, the tremor of the gas induced vortex motion. Just after the spherical body formation, planets started to revolve around the sun, which also rotates around itself, in a single plane. Based on the condensation point of heavy metals at 1atm, the temperature at the planetary formation is estimated to be around 5000 degree. The liquid drop of sun should be further cooled down to solidify. On the contrary, the surface temperature of the sun is measured to be more than 6000 degree, and it should be much hot inside. In other words, the sun should be heated up after its condensation in order to become a fireball with flares in coronas at two million degree. In this chapter, let us think about the genesis of the shining sun as we see everyday in the sky.

After the condensation of 99.9% of solar materials into a liquid droplet as the sun at around 5000 degrees at 1atm, the liquid sun should start to solidify gradually at around 3000 degrees from the surface by radiation cooling. At present, however, the sun burns to radiate an enormous energy corresponding to four thousand billions of nuclear power plants everyday. To tell the truth, the earth also produces geothermal energy for a spa and a volcano under the ground with much smaller scale than the sun. This energy is derived from the nuclear power of radioactive heavy metal elements enriched at the core of the earth. Since the mass of the sun is a million times larger than that of the earth, enormous amounts of radioactive materials should be enriched at the center of the sun. Since uranium is the natural heaviest element with the atomic number 92, almost all the uranium should be accumulated at the center of the solar system, that is, the core of the sun. This nuclear energy may increase the temperature of the sun so high from inside as geothermal energy of the earth, resulting in the genesis of the sun shining by a black body radiation. The solar flare or the volcanic eruption of the sun sometimes records around sub-billions degrees. Once 99.9% of the mass of the solar system accumulated into the sun, its great gravity produces a super-high pressure in its atmosphere around the sun, preventing the vaporization of the burning sun by itself. The nuclear burning in the sun started naturally after distillation, purification, and enrichment of the radioactive heavy metals such as uranium during the condensation. Since the sun is surrounded by some light elements such as a thick helium gas, the nuclear radioactive energy is transferred to those elements as a heat in coronas at the surface. It is as clear as a daylight that the radiation, heat, and materials produced by the sun is the main driving force for the solar system to perform the weather phenomena as described below.

Chapter Nine
Weather of the Solar System

The present solar system is organized as a kinetic system under the control of gravity of the sun. All the bodies revolving around the sun take elliptic orbits within the sphere of the gravity. Some bodies do not come back naturally to the system once they go out from the sphere of gravity while others sometimes fall into the sun within the sphere. Not only the revolution, the solar system is known to perform complex meteorological phenomena. In this chapter, we study the meteorology or general weather condition of the solar system together with a great circulation of materials.

The sun radiates energy, which travels everywhere in the solar system and also in the Cosmos. The sunshine blows the wind on the land, stirs up a vapor from the sea, makes clouds in the sky, and brews a hurricane and typhoon on the summer ocean. Almost all the weather phenomena are the general circulation of matters on the surface of earth, which is induced by the radiation energy from the sun. In the same way, matters in the solar system are generally circulated by the radiation energy from the sun. For example, a comet tail vapors intensely from the ice ball heated up by the radiation from the sun.

The vaporized water gas rises up in the solar system to the Oort cloud, where the vapor condenses and congeals to an ice body. The ice body gradually grows heavier and heavier by congelation of the water vapors rising from inside to outside of the solar system, and finally the ice body starts to fall down to the sun as a comet. This is the general circulation of water by comets in the solar system. So, the falling of icy comets and meteorites to the earth and the sun is a meteorological natural phenomenon just like a rain and a hailstone. In the same way, the surface of Uranus and Neptune is covered with ice by the condensation of water vapors. Physical conditions such as temperature, atmospheric pressure, and location of the nuclei for freezing determines the position, amount, and timing for condensation and freezing of water vapors into ice in the solar system. The sea could not be created from the element on the original earth, that is, an iron ball. The ocean sea may be a gift of welcome comets and meteorites from the heaven. On the other hand, there is no water on the moon, whose composition of rocks is quite similar to that on the earth. It is not so easy to explain this difference of the presence of water on the earth and the moon. Maybe it is because of the difference of total mass, gravity, and atmospheric composition of the earth and the moon. However, we could not prevent to imagine another romantic hypothesis that a big icy welcome comet fell down from the heaven to the earth once upon a time to create living things on the earth.

Now, we have discussed general circulation of water in the solar system from the observation of a comet tail. Is water the only material circulating in the system? That does not sound. It is quite likely that methane and ammonium also circulate generally in the system. Furthermore, vaporizable low molecular mass organic chemicals may circulate generally in the system and condensed around Mars. It is possible that we have an acidic rain of comets in the solar system.

While material circulation plays a central role in meteorology in the solar system, a part of materials are getting away from the system. It is no doubt that the solar system loses materials to some extent everyday because outer space is much more low pressure than the solar system. Probably, the solar system releases some materials constantly to outer space. On the other hand, the sun always supplies some materials to the system. The sun always produces not only enormous energy but also low atomic mass elements by nuclear fission. Among them, some elements evaporate at 6000 degrees on the surface of the sun. The gas at a high temperature under high pressure erupts as a solar flare. The released gas surrounds the sun forming an atmosphere, which shines as coronas. The gas also blows away from the sun into the solar system as a stream of transporting materials. Some materials released from the sun may condense on the surface of planets. In this way, the solar system performs various weather phenomena of material circulation such as supply of materials from the sun, general circulation in the solar system, and release of materials outside from the system.

Weather conditions play an indispensable role in creation and development of the solar system in its history. Especially, the general circulation of water, organic, and nitrogen compounds must play an essential role in the creation and growth of living things. So in the last chapter, let us think about the story of creation of life from a viewpoint of the weather in the solar system.

Chapter Ten
Comets and Creation of Life

A living thing is a self-reproducing catalyst surrounded by a biological membrane, which proliferates stably as an organism. As far as we know, the earth is the only planet, on which living things are confirmed to proliferate, in the solar system. When, where, and how were the living things created? Is the creation of life an accident or inevitable in history of the solar system? Is it usual or exception for planets to have living things in the solar system? Is it possible that some living things exist in other planets? And, are there any stars, whose planets harbor living things, in the milky Galaxy? The key to answer these questions should exist in the law of general circulation of elements essential for living things in the solar system.

Almost all the elements essential for life are relatively light elements whose atomic numbers are less than thirty, while some trace elements such as molybdenum and iodine have atomic numbers more than forty. Low atomic number elements accumulate in outer planets such as Uranium and Neptune, which are rich in water or an essential material for life. These planets, however, stay far away from the sun and thus almost freeze over without any sign of life creation. Other icy bodies also exist as comets in space outer than planets. In order to create life, these icy bodies should fall down to inner warm planets to melt with heat from the sun. Judged from the temperature, those planets should be Venus, Earth, and Mars. The surface temperature of Venus sometimes reaches four hundred degree, at which water vapors to become a steam. The temperature of surface of the earth is around 20 degrees, which satisfies the condition for life to be created. Mars also satisfies some of conditions for life. If the sunshine becomes a little bit stronger, Mars could create living things. Or it is possible for Mars to have created life long time ago. The earth and Mars are rich in iron, which could react with water and retain hydrogen and oxygen elements as some non-vaporizable form of iron oxide and iron hydroxide. The two planets also contain trace minerals essential for life such as magnesium and calcium. The two planets, however, do not contain main elements so much for living things, that is, hydrogen, oxygen, carbon, and nitrogen. The comets falling down from the heaven supply those elements essential for life, and thus it must be a welcome rain for living things on the earth.

When icy comets fall down to the rocky planets close to the sun by meteorological phenomena of the solar system, the creation of life is taking on a reality. Organic compound might be rich in asteroid belt, Jupiter, and

Saturn. Chemical evolutionists such as Oparin speculate that an inorganic vapor discharge created organic compounds during chemical evolution of the solar system. It is not clear whether the discharge occurred on the earth or in space of the solar system. However, since the sun radiates radioactive rays including a discharging beta-ray produced by nuclear fission to everywhere in the solar system, it is quite likely that some organic compounds could be created in space along with the general circulation of water and small chemical compounds. If icy meteorites, which have packed the synthesized organic compounds in them, fall down to the earth, the first creation of life might become reality somewhere on the earth.

The sunshine not only warms the solar system but also circulates materials generally and evolves chemical materials by electric discharge. After the welcome rain of comets and meteorites from the heaven to the ground, life might be created and grew dramatically as the first rain in decades on a desert grows and blooms grasses and flowers at once. The sun is the origin of energy, and comets might be the source of materials for life.

Reference Book

Chronological Scientific Tables (2007) by National
Astronomical Observatory (ed.), Maruzen Co., Ltd.
http://www.rikanenpyo.jp

Acknowledgements

Twinkling stars in the sky always tell us some romantic stories. The Big Dipper, Orion, and Cassiopeia shine brilliantly in the vast and boundless space. We have a key to open the secret of stars in the solar system. A vortex of the primary vapor in chaos gave a birth to spherical bodies as drops by condensation in the solar system. When the condensation and subsequent solidification completed, all the spherical bodies freed themself from the primary chaos and when they got into the clean space, the sun and moon started shining. After a great success of European science, now we can start to tell a great story of the genesis of the sun as a natural philosophy with modern technology and precise data with accuracy.

Development of natural science almost reaches to the secret of the solar system, and we not only understand the history of the sun but also succeed in developing a new energy plants, and the dream of space traveling is coming true. From now on, we need to step on a stage for establishing a natural law for development of space resources and for setting a principle to manage the solar system.

What is going on in the solar system now? And where shall we go in the future? Are there any living things outside the solar system? How was the primary vapor created in the Galaxy? It is my pleasure to have an opportunity to tell the story of birth of the Galaxy, which creates a lot of stars in the Milky Way.

We look at the sun rising up and setting down naturally every morning and every evening, respectively. The sun records profound truths and history. The sun warship started with human history. The sun is indispensable for agriculture together with soil and water. The ancient Egyptian prayed to five gods of the sun. The sun warship creates solar calendar, which became the basic culture of our modern life. The sun shines on the land everywhere to grow up living things with love.

Finally, I would like to express my special thanks to my family, Yoko, Kodai, Arisa, and Hiroto for supporting my life and writing this book.

At home in Kasukabe

Jun 18th, 2015

Hiroyuki Aizawa